Rising Tides

A Story of Survival in the Age of Flooding

By

Muhammad Edogi

Copyright © 2023 by Muhammad Edogi

No part of this book may be reproduced or stored in a retrieval system or transmitted in any form or any means of electronic, mechanical photocopying or otherwise without express written permission of the publisher

Table of Contents

1. Introduction 4
   - Rising Tides: Understanding the Science 8
     - Causes of sea level rise 9
     - Climate change and its effects on sea level 14
     - Solutions for Air Pollution and Climate Change 16
   - The Flooding Crisis 23
     - Economic and social consequences 26
     - Strategies for survival 29
     - Examples of successful adaptation measures 38
III. The Future of Flooding 39
   - Predictions and projections 41
     - The role of policy and technology 48
   - Conclusion 52
     - Call to action 54
H. References and Further Reading 55
     - Suggestions for additional reading 56
   - Facts about floods: 56
   - Types of Emergencies - Floods 58
   - How to Tell If Your Property Is in a Flood Zone 60
   - Are You At Risk of Flooding From Excess Rainfall? 62
   - Urban Risks - Flash Floods 66
   - What is a Flash Flood and How Do I Prepare For One? .. 69

1. Introduction

Introduction Flood Insurance protects your house & possessions from loss by rising water from the outside. Think about a river or creek overflowing into your home... a frightening thought. Homeowner's and other property insurance specifically exclude this peril.

If you own a house in a known flood risk area (i.e., the 100-year floodplain) with a bank loan, your mortgage bank will normally require flood insurance. For most homeowners, handling this mortgage bank flood insurance requirement is all they focus on and they ignore their true flood hazard. Then when a major storm does come, they have inadequate flood insurance coverage often with too little coverage on their house (often only the home loan balance) and no contents protection.

Also, over 25% of flood damage happens each year to properties outside of a known flood risk area (100-year floodplain). Central Texas had a recent example of an "out-of-the-blue" rain event that caused very intense flooding well beyond the known flood risk areas. The so-called "Marble Falls Rain Bomb" in June 2007 damaged over 100 homes & business around the city of Marble Falls with a very sudden 19 inch rainfall. A "Preferred Risk Flood Insurance Policy," available to homeowners beyond the 100-year floodplain, can protect your home and possessions at a very modest price.

My city of Austin is part of the Central Texas "Flash Flood Alley" and has a long history of major flooding along its creeks and the Colorado River. Dams located on Lake Travis and Lake Buchanan, built in the 1940's, has helped control the very destructive flooding of the Colorado River. Today, the biggest risk is along the many creeks in our urban areas and the Colorado River south of Lady Bird Lake dam. Shoal, Bull and Walnut Creek's in North Austin plus Onion and Williamson creeks in South Austin have considerable history of inundating adjacent areas.

Our neighboring Hill Country also has many creeks subject to flooding plus several major rivers that can rage with great torrents after heavy rain. The Llano and Pedernales Rivers both have had major flood events in recent years. The Llano River, surging into Lake LBJ has caused major flood damage along its normally calm waters on several occasions.

The hardest part of understand both your flood risk and flood insurance policies is the terminology. Most folks are confounded by its mix of insurance and engineering terms. Once you have a key to decipher the flood insurance nomenclature, things will make more sense. You also want to understand what your "Flood Zone" designation means. Finally, I have included an overview of the main components of a flood insurance policy.

Flood Insurance Terminology:

Base Flood Elevation - This is the level at which there is a 1% chance of flooding in any given year. A building that is located on land below the "Base Flood Elevation" is inside the 100-year floodplain.

Elevation Certificate - Clarifies the relative elevation of your house in relation to the know flood risk. This allows for more accurate rating of the flood insurance policy and may reduce your flood insurance rates.

Flood Maps ("FIRM" - Flood Insurance Ratings Maps) - Created by FEMA's (Federal Emergency Management Agency), these maps were created to determine which land areas are likely to be flooded. These maps are based on surveys of the elevation of land areas relative to known flood risks (creeks, rivers, lakes, etc.).

Floodplain - Any normally dry land area that is susceptible to being inundated by water often because it is adjacent to a watercourse. The 100-year Floodplain is the land that would be inundated by a 100-year flood event.

Flooding - Rising water from outside enters a structure. An example would be a house inundation from a flash flood. The flood peril also includes mudslide.

Hundred Year Flood - An engineering term used to describe the relative flooding risk. A house that is located inside the Hundred Year Floodplain is considered to have a 1% chance of being flooded in any given year. Most mortgages require that a house that is located in a Hundred Year Flood risk area must be insured for flood.

LOMA (Letter of Map Amendment) - Document used to establish that a building is not located in a Special Flood Hazard Area. A typical situation in which a LOMA would be important is when a part of a house lot is subject to flooding in a 100-year storm but the house itself has been built at a higher elevation.

National Flood Insurance Program - This is the government agency that provides insurance for the flood peril in the United States. Insurance companies are licensed to sell flood insurance policies for this government agency. All financial backing, rules and contract terms are set by the National Flood Insurance Program which is part of FEMA.

Special Flood Hazard Area - A geographic area that is prone to flooding. An example would be an area adjacent to a river that has an elevation low enough to be subject to flooding.

Flood Zones Designations:

- A - River / stream flood risk AE - River / stream flood risk with mapped base flood elevations
- AO - River / stream flood risk with shallow water depths (1-3 feet)
- AH - River / stream flood risk with shallow water paths (flows of 1-3 feet)
- V - Coastal or Storm Surge flood risk
- VE - Coastal or Storm Surge flood risk with mapped base flood elevations
- X - Not a Special Flood Risk Area (elevation above the 100-year floodplain)

Flood Insurance Overview

Property Coverage's:

Building - Provides protection up to your limit for damage or destruction of your house or other dwelling from peril of flood including rising water and mudslide.

Contents - Provides protection for your clothes, appliances, furniture and other possessions at your residence from peril of flood including rising water and mudslide. Flood Insurance offers "Actual Cash Value" as the basis of settlement. Contents coverage is optional and has a separate deductible.

Secondary Structures (fences, sheds, etc.) - None (No coverage is extended to secondary structures from the standard flood policy. Coverage is only available for the main structure.)

➢ Rising Tides: Understanding the Science

Rising Tides: Understanding the Science

Rising tides are a phenomenon that is becoming increasingly common as the earth's climate warms. As temperatures around the world rise, sea levels are rising at an unprecedented rate, resulting in the flooding of low-lying coastal areas. In order to better understand the science behind rising tides, it is important to explore the causes, effects, and potential solutions.

Causes of Rising Tides

The primary cause of rising tides is global warming. As the average temperature of the earth increases, more ice melts from glaciers and ice caps. This melted water adds to the volume of the ocean, resulting in higher sea levels. Additionally, as temperatures rise, the ocean water expands, leading to a further rise in sea levels.

Effects of Rising Tides

The effects of rising tides can be devastating. Low-lying coastal areas are often the most affected, as they are prone to flooding. This flooding can destroy homes and businesses, and can lead to displacement for those living in the area. Rising tides also pose a threat to marine ecosystems, as higher sea levels can lead to

Stalinization of coastal wetlands, damaging fish and other wildlife.

Solutions for Rising Tides

There are several potential solutions for rising tides. The most effective solution is to reduce global warming. This can be done by reducing emissions of greenhouse gases and switching to renewable energy sources. Additionally, coastal communities can invest in infrastructure that is resilient to flooding, such as seawalls and floodgates. Finally, communities can take steps to reduce the impacts of flooding, such as building higher roads and buildings.

Rising tides are a serious environmental issue, and one that requires serious action. Understanding the science behind rising tides is the first step in finding effective solutions to this problem. With the right strategies and the right actions, it is possible to protect coastal communities and ecosystems from the devastating effects of rising tides.

- ## Causes of sea level rise
  - ### Melting of ice caps and glaciers

Melting of ice caps and glaciers is a major contributor to sea level rise. As the Earth's temperature increases, due to rising greenhouse gas levels, ice at the poles begins to melt and eventually flow into the ocean. This results in a direct addition of water to the ocean, causing sea levels to rise. The melting of ice from Greenland and Antarctica is especially significant, as these are the largest ice masses on the planet and their melting would have a major impact on sea level.

  - ### Thermal Expansion of Ocean Water

Thermal expansion of ocean water is a phenomenon caused by the rise in global temperatures those results in sea levels rising. As the average temperature of the earth's atmosphere increases, more energy is absorbed by the ocean, resulting in the water expanding. This expansion causes the sea level to rise, leading to flooding of low-lying areas and the displacement of coastal communities.

I. **Causes of Thermal Expansion**

The primary cause of thermal expansion of ocean water is global warming. As the average temperature of the atmosphere rises, more energy is absorbed by the ocean, resulting in the water expanding. This expansion is greater in warmer waters, meaning that the ocean near the equator is more likely to expand than the ocean near the poles. Additionally, as the ocean warms, it expands and sea levels rise.

I. **Effects of Thermal Expansion**

The effects of thermal expansion can be devastating. Low-lying coastal areas are particularly vulnerable to flooding, as rising sea levels can lead to the displacement of entire communities. Additionally, rising sea levels can damage marine ecosystems, as saltwater can enter and damage wetlands. This can damage fish and other aquatic life, as well as the plants that live in the wetlands.

II. **Solutions for Thermal Expansion**

There are several potential solutions for thermal expansion. The most effective solution is to reduce global warming. This can be done by reducing emissions of greenhouse gases and switching to renewable energy sources. Additionally, coastal communities can invest in infrastructure that is resilient to flooding, such as seawalls and floodgates. Finally, communities can take steps to reduce the impacts of flooding, such as building higher roads and buildings.

Thermal expansion of ocean water is a serious environmental issue, and one that requires serious action. Understanding the science behind thermal expansion is the first step in finding effective solutions to this problem. With the right strategies and the right actions, it is possible to protect coastal communities and ecosystems from the devastating effects of rising sea levels.

Thermal expansion of ocean water is a physical phenomenon that occurs when the temperature of the ocean increases, causing the volume of the water to expand. The expansion of the ocean is due to the fact that the molecules in the water move faster and occupy more space when they are heated. As a result, the ocean water level rises.

Thermal expansion of ocean water is a significant contributor to sea level rise, which is one of the main impacts of global warming. As the planet's temperature continues to rise, the ocean absorbs more heat and its volume increases, leading to a corresponding rise in sea level.

There are several factors that contribute to thermal expansion of ocean water, including increased atmospheric temperatures, increased greenhouse gas emissions, and changes in ocean currents and circulation patterns. In addition, melting ice from glaciers and ice sheets also contributes to sea level rise by adding water to the ocean.

The effects of thermal expansion of ocean water can be seen in coastal communities around the world, where rising sea levels can lead to flooding, erosion, and the displacement of people. In addition, rising sea levels can also have significant impacts on marine ecosystems and wildlife, as well as on the infrastructure and economies of coastal communities.

To mitigate the effects of thermal expansion of ocean water and sea level rise, it is important to reduce greenhouse gas emissions and implement strategies to adapt to the impacts of global warming. This may include investing in coastal protection and flood management infrastructure, as well as implementing sustainable land-use practices and reducing other forms of environmental degradation.

> **Increased greenhouse gas emissions**

Global warming is one of the most concerning environmental issues of our time, and one of the main contributors to it is increased greenhouse gas emissions. As greenhouse gas emissions rise, the atmosphere and oceans absorb more heat, resulting in a warming of the planet. This warming has been linked to an increase in sea level, as more ice melts from glaciers and ice caps, and as ocean water expands. Additionally, global warming causes more extreme weather events, such as hurricanes, which can lead to an increase in sea level.

The main cause of increased greenhouse gas emissions is the burning of fossil fuels, such as coal and oil, for energy production. Other sources of greenhouse gas emissions include agriculture, deforestation, and the manufacture of certain products. All of these activities release large amounts of carbon dioxide and other gases into the atmosphere, leading to a warming of the planet.

Reducing greenhouse gas emissions is essential for preventing sea level rise. This can be done by using renewable energy sources, such as solar and wind power, instead of fossil fuels. Additionally, reducing deforestation and investing in sustainable agriculture can help to reduce emissions. Finally, governments and businesses can take action to reduce emissions by implementing policies and regulations that limit emissions.

Sea level rise is a serious environmental issue that is caused by increased greenhouse gas emissions. In order to prevent sea level rise, it is necessary to reduce emissions from all sources. By taking action to reduce emissions, we can protect our coastal communities and ecosystems from the devastating effects of rising tides.

➤ Land-based ice loss from Greenland and Antarctica

Land-based ice loss from Greenland and Antarctica is one of the main contributors to rising sea levels. As global temperatures rise, the large ice sheets that cover Greenland and Antarctica are melting at unprecedented rates. This melting ice is causing the sea level to rise at an alarming rate, threatening coastal communities and ecosystems.

In Greenland, the majority of the melting ice is caused by calving glaciers. Calving glaciers are large icebergs that break off from the ice sheet and enter the ocean. As these icebergs melt, they add large amounts of water to the ocean, causing sea levels to rise. In Antarctica, melting ice is caused by both calving glaciers and surface melting of the ice sheet. As the surface of the ice sheet melts, the water runs into the ocean, causing sea levels to rise.

The effects of land-based ice loss from Greenland and Antarctica are devastating. Rising sea levels can lead to the displacement of coastal communities and can damage marine ecosystems. Additionally, melting ice can cause changes in ocean salinity, which can affect the food chain and lead to an increase in ocean acidification.

In order to prevent land-based ice loss from Greenland and Antarctica, it is essential to reduce global warming. This can be done by reducing emissions of greenhouse gases, such as carbon dioxide and methane, and switching to renewable energy sources. Additionally, governments and businesses can take action to reduce emissions by implementing policies and regulations that limit emissions.

Land-based ice loss from Greenland and Antarctica is a serious environmental issue with devastating effects. In order to protect coastal communities and ecosystems, it is essential to reduce global warming and to take steps to reduce emissions. With the right strategies and the right actions, it is possible to reduce the impacts of melting ice and to protect our planet from rising sea levels.

- ## Climate change and its effects on sea level

Turning down the Heat: Why a 4 Celsius World Must Be Avoided is a MOOC course that this page aims to summarize.

The 2013 Intergovernmental Panel on Climate Change report and the Turn down the Heat report serve as the article's primary sources of information (IPCC).

The primary objective of this article is to disseminate knowledge about climate change and its effects on various aspects of human life, including risks to food production and agriculture, risks to the world's water resources, impacts on ecosystems and biodiversity, consequences of sea level rise, and risks to human health.

Some viewpoints will be explored in light of perceptions developed through discussions, comments made during exercises, as well as through lectures and reports present in this Climate Change MOOC course, which is extremely intriguing.

By reading this essay, students and the general public will understand how urgent a change in human behavior is required to stop future generations from suffering the negative effects of a world that is 4 degrees Celsius warmer.

The most recent IPCC assessment states that it is "at least 95% likely" that human actions, particularly the combustion of fossil fuels, are to blame for the global warming that has occurred since the 1950s. This quotation exemplifies the primary issue now being discussed in the most recent climate change reports. due to human behavior, including excessive use of unsustainable manufacturing methods and fossil fuels in the primary means of transportation. The fact that the world has been getting much hotter is generally acknowledged. The majority of scientists have stated that a global temperature increase of 2 Celsius is unavoidable. However, the human race still has time to act quickly to prevent a drastic rise in global temperatures.

According to reports like the ones described above, an increase of 4 degrees Celsius must be avoided at all costs to ensure the safety of future generations. If human behavior does not change, there will be severe effects on important industries, including: Agriculture (negative effects on fisheries, livestock, and crops); Water Resources (changes in rainfall patterns, decreased availability of fresh water, increased salinity of water); Ecosystems; and Human Health (negative effects on landscapes, altered biomes, intensified droughts and floods, and increased spread of diseases like cholera and malaria as a result of the increase in salty water).

The majority of the effects of climate change have already begun to be seen in a number of places throughout the world, including South Asia, Africa, South America, Antarctica, and Greenland. Global sea level rise, an ongoing rise in green gas emissions, an increase in heat waves in various parts of the world, and acidification of the oceans are all directly tied to the effects of climate change. It is common knowledge that these effects have had varying degrees and types of impacts on various parts of the world.

Brazil serves as a solid example of how to put the concepts mentioned in this article into practice by focusing on a country analysis of the effects and repercussions previously described. Brazil is a huge nation with a continental footprint, as is well knowledge. As a result, it is expected that different Brazilian regions would experience varied effects of climate change. This will occur since each location has a distinct environmental biome, and as a result, each region's productivity and economy will change. The Human Health industry would probably be the one most impacted by changes in global temperatures in Brazil out of all the major industries mentioned earlier.

A serious impact on not only human health as it is typically understood, but also on the standard of living in these communities, may be seen. Brazil's health system has had significant issues for at least the last three decades. Brazil has been under investing on health issues, and this is a fact that transcends all theories and data. Brazil will undoubtedly face a significant task to overcome if you combine this historical issue with the health management system with the adverse effects that Climate Change has been raising globally. Brazil's most pressing issue would be this important sector.

This is connected to not just the problems that have already been discussed, but also to a critical sector where a cascade effect is more pronounced. The health and way of life of the inhabitants are directly impacted by factors impacting agriculture, water management resources, and ecosystem balance. The population's health will suffer if they do not have a sufficient supply of resources for food, water, and sanitation. The Northeast and North of Brazil would be the most affected regions, and because Brazil's health system is not as effective as it should be, an increase in diseases brought on by more pests and parasites would be a more difficult problem for Brazilian society to overcome.

- ## Solutions for Air Pollution and Climate Change

Air Pollution and Climate Change

What are the problems with air pollution and is it connected to climate change? There has been some confusion about climate change. Here is an analysis of the evidence for climate change and its consequences with some possible actions that could be taken to reduce the damage of climate change.

In reality, there are two problems with air pollution. The simplest is air pollution by trace gasses. Here, small amounts of dangerous gasses (usually acids) are released in a chemical reaction, usually combustion. These gasses have a bad impact on the environment and must be eliminated. A good example is sulfur in coal. The sulfur in coal is oxidized by combustion inside a power plant and is washed out of the atmosphere by rain, making "acid rain". When enough acid rain is formed, it starts killing plants and fish. These pollution problems are readily traced and are usually not controversial. What is controversial is how to get rid of the pollution. Usually a procedure can be found, but it may be expensive. This problem will not be addressed further here.

The more complex problem is air pollution that causes a composition change in the atmosphere. This is exemplified by the increase in atmospheric carbon dioxide and methane and its impact on earth's average temperature. The theory and the best data indicate that if too much carbon dioxide and methane (greenhouse gasses) get into the air, they capture the visible radiation, hold the infrared radiation and change the earth's heat balance. This raises the average temperature of the earth's atmosphere, and so it is called global warming.

One thing that makes this theory controversial is that all fossil fuels generate carbon dioxide when burned, and the vast majority of our energy is obtained by burning fossil fuels, so it is very difficult (and expensive) to reduce the amount of carbon dioxide that is emitted. Therefore, there is a very strong motive to disbelieve this theory.

Another problem is that the earth has climate zones that move with average temperature, so the zone position changes as the average temperature increases. Thus at any earth position, the temperature may be increasing (due to global warming) or decreasing (due to zone position movement). Critics ask which trend they should believe. The answer, of course, is that it is the average of the temperatures in all climate zones that determines the average earth temperature. This average cannot be determined by a measurement in only one earth position and so it is still being argued.

A third thing that makes this global warming controversial is the impact it may have on the earth's livability. It may not be possible to just wait for the effects to become clear and then take action. We may have to decide on an action plan now.

If indeed the earth is warming, then several things will happen:

- The earth's glaciers and ice caps will be reduced, and eventually disappear. Less of the visible radiation on the earth will be reflected into space, and more will be captured which will tend to increase the earth's average temperature. Also, some of the ocean's most productive zones are under ice, so loss of ice may result in a loss in the ocean's fish production.

- The melted ice will raise the mean sea level and low-lying land will be submerged. If so, some of the most important and valuable real estate in the world will be submerged.

- The climate zones will move north in the northern hemisphere (or south in the southern hemisphere) and some old fertile agriculture zones will dry up and some new zones will be waterlogged. The consensus is that there will be a net loss of agricultural area.

- The oceans will warm and spread. This will kill many reefs in the ocean and cypress forests on the edge of the ocean where fish breed with a resulting loss of fish production. Hurricanes will also increase in strength.

- The aerosols in the earth's atmosphere (fog, dust, ice particles, sulfur dioxide, etc) will change. An increase will increase the amount of visible radiation reflected by the atmosphere, and this could decrease the amount of radiation absorbed by the atmosphere. Most experts expect an aerosol increase and a resultant reduction in solar absorption.

- Most important, the permafrost in the arctic is expected to melt. This will cause the vegetation frozen in this layer to decompose and emit methane and carbon dioxide that would raise the temperature more. Thus the warming caused by man would cause more warming caused by nature.

Until now, mankind put carbon dioxide and methane into the

atmosphere, and warming started. If stopped a new equilibrium would form and the warming would stop, but at a higher average temperature. There are processes that absorb the new carbon dioxide and aid the formation of this equilibrium. Two of the most important of these processes are forest growth and carbonate rock formation by plankton. Clearly mankind is overwhelming these processes, because the carbon dioxide content of the atmosphere is growing rapidly. Part of this problem is that mankind is cutting the forests, but largest part of the problem is the greenhouse gasses from fossil fuels.

In the future, if mankind reduces his carbon dioxide contribution enough to drop below the natural absorption capability, there will be at least two warming processes that still grow-the loss of the reflectivity of ice at the poles, and the carbon dioxide and methane production by decomposing permafrost vegetation. If the effect of these processes rises above that of the natural absorption processes, the warming trend will continue without mankind's contribution. This automatic temperature increase is called runaway warming. The only way to stop runaway warming is to provide a new means of removing carbon dioxide from the atmosphere.

The Evidence

The evidence shows the following trends:

- Some areas show a warming trend and some show cooling. A computer model is required to interpret the data because climate zones are shifting as well as warming. Generally, however, warming trends seem to dominate.

- The glaciers and ice caps are melting.

- The melted ice does appear to be raising the mean sea level; although this measurement is more controversial. The mean sea level appears different at different earth positions.

- The climate zones are moving north in the northern hemisphere (and south in the southern hemisphere). This results in desertification in some productive agricultural areas and water logging in others.

- The oceans are warming and some reefs are dying and some cypress forests are dying and hurricanes appear to be increasing in strength.

- Aerosols in the earth's atmosphere are changing, but they are hard to measure. New, more accurate measuring devices are just coming on line.

- The permafrost is melting, especially in northern Canada and Siberia.

Several computer programs that integrate these measurements exist and they are being tested. They show a climate-warming trend, but the earth does not appear to have reached runaway. The accuracy of these programs is not yet completely confirmed with data, but this accuracy is improving.

Is Action Required?

Many still do not believe in climate warming. A counter theory has been proposed. This theory says that the warming trend that we observe is due to changes in solar radiation level and earth rotation axis wobble. Since there is nothing we can do about these causes, these critics propose that we do nothing that would upset the world economy, and wait to see what happens. This procedure could be very dangerous, as we shall see.

Suppose nothing is done. Then the following long-term bad effects are likely.

- The glaciers are part of the earth's fresh water storage system. If they disappear, the rivers will tend to flood in the winter and spring and dry up in the summer.

- The ice shelves in the Arctic and Antarctic are excellent fish food producers and if they disappear, this food source for fish may disappear.

- If the mean sea level rises to its maximum, some of the most important and valuable seacoast real estate in the world will be submerged.

- If climate zones move north (and south), significant numbers of productive agricultural areas will be lost.

- If the oceans warm to the maximum, a large portion of the earth's reefs will die, and many cypress forests will be damaged. This will damage the associated fish breeding grounds. These problems will cause the reduction of an important food source.

The above effects have, for the most part, a limited bad impact on the earth's livability. However, one impact of this cascade of events, the loss of reflective ice and the melting and decay of the permafrost, may cause runaway warming. Inaction would lock in all these bad effects and open us to many future problems that are even worse. If runaway is possible and something can be done, action now is absolutely required.

The Solution

A practical and economically positive solution may be possible without damaging the global economic system. This consists of an energy generating system that can reduce carbon dioxide emission and sequester the remainder. Specifically:

- A. Conservation, which would consist of substituting for fossil fuel power plants:
- B. Nuclear power plants where economical and safe
- C. Deep thermal well power plants where economical
- D. Ocean based wind and wave generators and solar cells to provide both base load energy and portable fuels.
- E. Electrical cars with solar cells to extend range.
- F. Alcohol and oil from waste wood, algae and kelp for portable power plant operations such as aircraft, trains, cars and trucks.
- G. Sequestering, which would consist of putting the carbon dioxide in the:
- H. Deep rock formations by use of deep thermal wells.
- I. Deep oceans by freezing the carbon dioxide and sinking it below the thermo cline.

There is a major problem with sequestering, however. Both the deep thermal well and the ocean based wind and wave generators are being developed by small companies that, under normal development procedures, would not be expected to have a large impact for 30 years, and would not be expected to start reversing the warming trend for 40 to 50 years. Now the ice caps and the permafrost layers are expected to melt in 15 to 25 years. Thus we may be in a state of runaway global warming before the solution can come on line. Timing may be important.

Timing and the Overall Capability

This situation contrasts with other green energy producers such as solar cells. Solar cells have a serious production limit caused by a shortage of both worker skills and refined solar cell materials, and could not ramp up into the dominant energy producer in a timely fashion.

A useful addition to the solution would be a process to sequester carbon dioxide from the atmosphere; however there are no currently available commercially viable ways to do this.

➤ The Flooding Crisis

Flooding is an increasingly common problem all around the world, and one that threatens both lives and property. Flooding is caused by a variety of factors, including changes in climate, increased development, and construction of dams or other infrastructure. As a result of these factors, floodwaters can quickly overwhelm areas, leading to significant damage and loss of life.

Climate change is one of the main contributors to the flooding crisis. As global temperatures rise, the amount of water in the atmosphere increases, leading to more intense rainstorms and flooding. Additionally, melting ice caps and glaciers can lead to an increase in sea level, making coastal areas more vulnerable to flooding.

Development and construction can also contribute to flooding. As development increases, less ground is available to absorb excess water, leading to increased runoff into rivers and streams. Additionally, the construction of dams can change the flow of rivers, leading to increased flooding downstream.

In order to combat the flooding crisis, it is essential to reduce emissions of greenhouse gases and to switch to renewable energy sources. Additionally, governments and businesses can take action to reduce development and to implement policies that protect natural areas from development. Finally, countries should work together to create and implement policies and regulations that limit the construction of dams.

The flooding crisis is a serious environmental and humanitarian issue with far-reaching consequences. In order to protect lives and property, it is essential to take action to reduce emissions, to protect natural areas from development, and to limit the construction of dams. With the right strategies and the right actions, it is possible to reduce the impacts of flooding and to protect our planet from this growing threat.

It is not only homeowners actually in foreclosure who are dealing with the fallout effects of the housing market meltdown. Although they are the hardest hit, many more people who never miss a mortgage payment also must face the consequences of the high foreclosure rates, and local governments must also deal with trying to keep order in the community with fewer resources and rising costs.

The list of groups and institutions feeling the pain of the foreclosure crisis encompasses nearly every aspect of American life. County governments are faced with budget crises as property tax revenue falls. Mortgage companies are so far behind on selling homes that have been foreclosed on that they are holding off on beginning more foreclosure proceedings. Local court systems could not handle more foreclosure lawsuits anyway, as they are also far behind.

And the group most affected by the foreclosure rates is the homeowners and neighbors who have large numbers of abandoned properties sitting in the community. Ripe for vandalism and squatting, these empty properties are magnets for crime. With fewer homeowners in an area even to report the disturbances, and police departments reeling from lower budgets and higher fuel costs, the suburbs may quickly turn into the new slums.

It should also come as no surprise that homes that have been abandoned or are not sold are falling into disrepair as no upkeep is done on them by the owners (usually the foreclosing bank). Broken windows are not fixed, leaky roofs lead to severe water damage, and a broken sump pump may cause a basement flood to remain undiscovered for months. These are far worse than just an overgrown lawn and will contribute to longer-term decline in house prices.

One good idea that local governments have had in some ares has been to take over these abandoned properties and make sure they are maintained or simply tear them down to improve the neighborhood. Although governments owning large portions of a community may not be a good idea, it can only be better than multinational banks owning these properties and leaving them to disrepair and crime.

But even governments that take over foreclosed homes must put the cost burden onto current homeowners through higher property taxes on fewer homeowners. In turn, the higher tax rates means that housing costs will be higher in particular areas, so home values must come down even further to induce buyers to move into an area. Falling property values, however, make it more difficult for homeowners to stop foreclosure by selling, and contribute to higher foreclosure rates.

Homeowners have also been forced into the mortgage trap as the housing market cools considerably. Even if they have never missed a payment, many home loans of the past decade were made up to 100% of the value of a property that may now have fallen 40% or more. Homeowners who wish to sell and upgrade or take a new job opportunity in another area of the country are unable to sell for enough to pay off their mortgage, whether they are in foreclosure or not.

Unfortunately, even as homeowners in foreclosure are forced to deal with the loss of their home and the negative consequences of this, the people institutions that are left behind often fare no better. The foreclosure crisis is contributing to a cycle of lower property values, more crime, and more desperate actions by local governments. Nothing yet, however, has been able to slow down the crisis in some areas of the country, and it may be that the mistakes of the subprime criminals and suburban sprawl will have to be paid for by mostly innocent **Impact on coastal communities**

• Economic and social consequences

Flooding can have significant economic and social consequences. Some of these consequences include:

Economic Consequences:

1. Property damage: Floods can seriously harm infrastructure, houses, and buildings.

2. Business disruption: Floods can cause a loss of revenue and disrupt enterprises.

3. Agriculture losses: Floods can harm livestock and crops, costing farmers money.

4. Cost of products and services rising: Because of disruptions to supply and transportation networks, floods can raise the price of goods and services.

Social Implications

1. Floods can compel people to leave their houses, which can result in the destruction of property and eviction of residents.

2. Health risks: Water-borne illnesses like cholera and dysentery are much more likely by floods.

3. Psychological distress: Communities affected by floods may experience stress, worry, and sadness.

4. Communities' resources, such as those for food, shelter, and healthcare, may be strained as a result of floods.

Flooding is a major problem that can have devastating economic and social consequences. Floodwaters can quickly overwhelm areas, leading to significant damage to buildings and infrastructure and disruption of commerce. Additionally, flooding can cause serious health risks for those in affected areas, leading to illness, injury, and even death.

The economic consequences of flooding can be significant. When buildings and infrastructure are damaged, it can be costly to repair or replace them. Additionally, flooding can disrupt businesses and industries, leading to a loss of income and jobs. Finally, flooding can cause damage to crops and livestock, leading to food shortages and famine.

The social consequences of flooding can also be severe. In some cases, flooding can displace entire communities, leading to loss of homes and disruption of lives. Additionally, flooding can cause health risks, leading to illness, injury, and even death. Finally, floods can lead to social unrest, as people struggle to deal with the consequences of the disaster.

In order to reduce the economic and social consequences of flooding, it is essential to take action to reduce emissions of greenhouse gases and to switch to renewable energy sources.

Additionally, governments and businesses can take action to reduce development and to implement policies that protect natural areas from development. Finally, countries should work together to create and implement policies and regulations that limit the construction of dams.

The economic and social consequences of flooding are severe and far-reaching. In order to reduce the impacts of flooding, it is essential to take action to reduce emissions and to protect natural areas from development. With the right strategies and the right actions, it is possible to reduce the impacts of flooding and to protect our planet from this growing threat.

- **Strategies for survival**

Flooding is something that every country in the world is familiar with. This is why basic flood survival should be known by everyone. Indeed, it has been a problem that we keep on seeing close to us, or in some far off place. We see a lot of news on flooding disasters that have claimed lives and property. Due to the hydrological cycle, we can't seem to get rid of all of the water as it is being recycled through rain, evaporation and storms.

That doesn't mean we just have to succumb to the disaster. Ironically, flooding is one of the disasters that are easiest to manage, given a proper disaster preparedness program.

Floods can be classified into two categories; the regular flood from sustained rainfall over long periods of time and the flash flood. The regular flood is what we normally encounter. It happens when heavy rains run for a long duration. The excess water that cannot be readily accommodated by the soil or storm drains causes the flooding. It steadily rises until it reaches disaster levels.

The second type of flood, the flash flood is more dangerous than regular floods. The fast rate at which the water level rises leaves people no time for safety measures. Some flash floods could happen in just minutes, catching people off guard. Flash floods happen in typhoons Bering heavy rains. During these times, it is advisable to

always have at least one member of the family awake and monitor the water level during the night. You can always catch up on sleep later. Flash floods have been reputed to reach areas where no flooding has ever happened.

Safety Measures during Floods

Minor flooding can be a nuisance. It destroys property and the clean up process is extremely tiresome. Major floods are disastrous. They can claim lives and cause immense damage to property. If you are planning to buy a house or relocate to a new neighborhood, it would be prudent to ask around the area about their flooding experiences.

As much as possible, stay away from areas the have been flooded before. It doesn't matter if the last flood occurred five or ten years ago, the fact that it has been flooded before indicates that it can be flooded again. The flooding indicates its low geographical topography and it's only a matter of time before another flooding occurs.

Unless there have been significant preventive measures implemented to prevent flash floods as well as regular flooding, you cannot safely conclude that flooding won't occur in the neighborhood again.

If you are living in a multi-story house, move important items like documents and fragile personal effects to the upper floors during heavy rains and storms. If possible, move all non-washable items like sofas and mattresses off the first floor.

Avoid driving through a flooded area. Most cars start to stall in six inches of water. It is very dangerous to get trapped in a car during flooding. You can get carried away in the current and washed away to deeper waters.

Basic utilities like electricity and tap water services may be cut off during flooding disasters to prevent other accidents like electrocution and disease from happening as a result of sewage over-flow. Part of your disaster preparedness program is to provide enough drinking water and lighting provisions like LED lamps, candles and matches to last a minimum of 72 hours after the danger has passed.

Prepare an emergency kit

An emergency kit for floods should include:

1. Basic supplies: food, water, first aid kit, battery-powered radio, flashlights, and extra batteries

2. Personal items: identification, cash, medications, important documents (passport, insurance)

3. Clothing and bedding: extra clothes, sturdy shoes, blanket, and sleeping bag

4. Tools: manual can opener, wrench to turn off utilities, shovel, and pliers

5. Sanitation: soap, toothbrush, towels, and toilet paper

6. Special items: infant formula, pet food, and important equipment for people with disabilities

7. Keep the kit in a waterproof container and store it in an easily accessible location. Update it regularly.

Elevate valuable items and electrical equipment

Elevating valuable items and electrical equipment is an important step in protecting your home or business from the dangers of flooding. Elevating these items will reduce the chances of them being damaged or destroyed in a flood. It is important to note that different items will require different levels of protection, so it is important to assess the specific needs of your items before attempting to elevate them.

For items such as furniture and other belongings, elevating them at least a foot above the ground can be an effective way to protect them. Additionally, it is important to make sure the items are securely placed and unlikely to move in the event of a flood.

For electrical equipment, it is important to take additional precautions. This may include elevating equipment at least one foot above the ground, as well as taking measures to protect it from water damage. Additionally, it is important to ensure that the equipment is securely placed and unlikely to move in the event of a flood.

Finally, it is important to ensure that the elevation is maintained during any heavy rains. Additionally, it is important to inspect the elevation regularly to ensure that it is secure and that the items are still safely elevated.

Elevating valuable items and electrical equipment is a necessary step in protecting them from the dangers of flooding. It is important to assess the specific needs of the items and to take the appropriate steps to ensure that they are securely elevated and unlikely to move in the event of a flood. With the right measures and the right precautions, it is possible to ensure that your valuable items and electrical equipment are safe

Before a flood, elevating priceless possessions and expensive electrical equipment can help prevent damage and reduce losses. Following are the steps:

A. Transfer critical papers, gadgets, and furnishings to higher floors or vantage points.

B. Elevate electrical, HVAC, and appliance systems or install flood vents.

C. To keep water from entering the building, surround its perimeter with sandbags or other barriers.

D. To get rid of the standing water in your basement, think about installing a sump pump.

E. Before leaving, unplug all electrical devices and appliances.

F. Store potentially flooded areas above the level of hazardous substances like chemicals and pesticides.

G. To aid with recovery operations, mark the elevation level of your water main, gas meter, and electrical panel.

Make a evacuation plan

An evacuation plan is an essential tool for preparing for an emergency situation. It outlines the steps that need to be taken in order to ensure the safety of people in the event of a disaster, such as a fire, flood, or tornado. The plan should include information on evacuation routes, emergency shelters, and necessary supplies, as well as communication plans and evacuation procedures.

When creating an evacuation plan, it is important to include the following elements:

Establish Evacuation Routes: Establish multiple evacuation routes away from the area of danger, and identify possible safe destinations.

Plan for Emergency Shelters: Identify locations that can provide emergency shelter in the event of an evacuation.

Prepare a List of Necessary Supplies: Make a list of necessary items, such as food, water, medical supplies, and communication devices, that should be taken if an evacuation is necessary.

Develop Communications Plans: Establish plans for communicating with family and other important contacts in the event of an evacuation.

Practice Evacuation Procedures: Make sure everyone is familiar with the evacuation plan and knows what to do in the event of an emergency.

By creating an evacuation plan, families and businesses can be better prepared in the event of an emergency. Taking the time to plan for an evacuation can help ensure the safety of everyone involved.

- Creating an evacuation plan can help ensure a safe and orderly departure in case of a flood. Steps to consider when making a plan include:
- Designate a meeting place for all family members to gather in case you get separated.
- Plan multiple evacuation routes to avoid flooded areas.
- Make arrangements for transportation, including identifying potential shelters, hotels, or friends and relatives to stay with.
- Plan for pets and other animals, including securing food and water, leashes, carriers, and any necessary documents.

- Create a list of important phone numbers and email addresses and store it in a waterproof and easily accessible location.

- Identify potential hazards in your home and surrounding area, such as gas lines and electrical equipment, and know how to turn them off.

- Review the plan with all household members and practice evacuation drills.

Pack an emergency supply kit and keep it in a ready-to-go location.

Staying informed about flood warnings is essential for protecting yourself and your property from the dangers of flooding. The National Weather Service issues various flood products, including flash flood warnings, flash flood watches, river flood warnings, and urban and small stream flood advisories [1]. It is important to stay aware of these warnings and to take appropriate precautions if your area is at risk of flooding.

It is also important to stay up to date with the latest weather forecasts and warnings. The NWS provides official warnings, watches

Staying Informed of Flood Warnings

Staying informed of flood warnings is an important step in protecting yourself and your property from flooding. The National Weather Service (NWS) issues various flooding products, including flash flood warnings, flash flood watches, and river flood warnings

[1]. These warnings are designed to keep people informed about the potential for flooding and any other related risks. It is important to pay attention to these warnings so that you can take the necessary precautions to protect yourself and your property.

One of the best ways to stay informed of flood warnings is to listen to NOAA Weather Radio or watch local television

[2]. These sources will provide up-to-date information about any flooding that is occurring or that is likely to occur. Additionally, it is important to stay informed about the weather in your area so that you can be aware of any weather systems that could lead to flooding

[3]. Staying informed of flood warnings is an important step in protecting yourself and your property from flooding. By taking the time to stay informed of any warnings, you can be better prepared to take the necessary steps to protect yourself and your property.

Staying informed of flood warnings is crucial in order to take necessary precautions and stay safe. Steps to follow include:

- A. Monitor local news and weather reports for updates on flood conditions.
- B. Sign up for emergency alerts and notifications from local authorities and weather agencies.
- C. Follow the guidance of local officials, including evacuation orders and road closures.
- D. Keep a battery-powered radio and extra batteries on hand to stay informed during power outages.

E. Download a weather app or use websites like the National Weather Service to track flood warnings in your area.

F. Have a plan to receive information during a power outage, such as a battery-powered or hand-crank radio.

G. Stay vigilant and be ready to act quickly if a flood warning is issued.

Keeping Sandbags on Hand

Having sandbags on hand is a great way to prepare for potential flooding. Sandbags can be used to contain and divert water away from homes and other structures, helping to protect them from flooding

[1]. Sandbags are also useful for protecting businesses, vehicles, and other property from flooding.

To prepare for potential flooding, it is a good idea to keep several sandbags on hand. Sandbags can be bought at most home improvement stores and some hardware stores. Additionally, many cities and towns have sandbagging programs where sandbags can be obtained for free or for a nominal fee

[2]. When storing sandbags, make sure to do so in a dry, well-ventilated area. Sandbags should also be stored in a place that is easily accessible, in case they need to be used in an emergency situation.

Having sandbags on hand is a great way to prepare for potential flooding. By taking the time to keep a few sandbags stored in a dry, well-ventilated area, you can be better prepared to protect your home and other property from flooding.

Avoiding Driving in Flooded Areas

Driving in flooded areas can be extremely dangerous and can result in serious damage to your vehicle, as well as a potential injury or death. It is important to take the necessary precautions to avoid driving in flooded areas whenever possible, as doing so can help to reduce the risk of damage to both your vehicle and yourself.

The first step to avoiding driving in flooded areas is to stay informed of any flooding warnings that have been issued

By paying attention to any flood warnings that have been issued, you can be better informed about the potential for flooding in your area and can avoid driving in any flooded areas.

Additionally, it is important to pay attention to the conditions of the roads that you are driving on. If you come across a road that is flooded, it is important to turn around and find an alternate route. Do not attempt to drive through flooded roads, as doing so can cause your vehicle to become stuck and you could be swept away by the water

By taking the time to stay informed of any flooding warnings in your area and by paying attention to the conditions of the roads that you are driving on, you can help to avoid driving in flooded areas and reduce the risk of damage to both your vehicle and yourself.

Now for tips on avoiding driving through flooded areas, see:

- Prior to driving, check the weather and road conditions.

- Driving into running water should be avoided, regardless of depth.

- If the road is flooded, detour and take a different path.

- If your car starts to float, get out right away and head for higher ground.

- If there are power lines, electrical poles, or fire hydrants in standing water nearby, avoid driving through those areas.

- **Examples of successful adaptation measures**

Flooding can have devastating consequences for communities, resulting in extensive damage to both property and lives. To help mitigate the effects of flooding, various adaptations measures can be taken. Here are some examples of successful adaptation measures for flooding that can help reduce the risk of damage.

The first adaptation measure is to develop better forecasting and early warning systems for flooding. This can involve improving monitoring systems and utilizing technology to better predict and detect flooding, allowing communities to be better prepared

[1]. Another adaptation measure is to create land-use policies that are designed to reduce the risk of flooding. Restrictions can be put in place on building in flood plains and other high-risk areas that are prone to flooding. Additionally, flood-resistant building designs can be encouraged, such as green roofs and elevating buildings to reduce the risk of flooding

[2]. Other successful adaptation measures for flooding include creating buffers and dams, planting vegetation, and restoring wetlands. These measures can help to reduce the risk of flooding and to mitigate the effects of flooding

[3]. By implementing these successful adaptation measures for flooding, communities can reduce the risk of damage and be better prepared for flooding.

III. The Future of Flooding

Flood protection is vital for any home at risk of flooding and in this article we'll discuss the different types of flooding and the crucial part the humble sandbag plays in flood defenses.

Flooding is becoming an increasing risk for many home-owners. In England and Wales alone, nearly 5 million households are at risk of flooding and as more and more homes are being built in high risk areas, the problem is expected to get worse.

Flooding is not just a problem for those living near the coastline or rivers either. Floods can happen anywhere and at anytime, particularly after a heavy rainfall and it is important to be aware of the risks of flooding and types flood protection available. There are numerous causes of flooding many of which can happen anywhere even if you live miles from a main source of water. The main causes of flooding are:

River flooding - rivers can burst their banks after heavy rainfall and this can lead to one of the most widespread types of flooding and reservoirs and lakes can burst their banks too.

Coastal flooding - is often caused by storms but with climate change expecting to cause rising sea-levels this type of flooding could increase.

Groundwater Floods - caused by lack of drainage of groundwater particularly in areas built on permeable rock

Surface and sewer water flooding - Caused by blocked drains and sewers, this type of flooding can hit any home and at anytime.

The UK's Environment Agency produces a map of areas deemed as high risk and it is vital that if you live in one of these areas you put some thought into proper planning to prevent flooding and to prepare for it by thinking about flood defenses and flood protection.

The aim of flood defenses is to: prevent, delay and limit the damage of water getting onto a home.

One of the most common and useful flood defense tools is the humble sandbag. These can be stacked into walls to prevent water from getting in and have been used for centuries in preventing flood risks. They are cheap, easy to store and provide excellent defense against flooding.

Unfortunately there are downsides to the sandbag. Traditional sandbags have to be filled-up in advance of any flooding. Unfortunately not every flood can be predicted and often there is not enough time to prepare the sandbags in advance.

However, a new type of instant sandbag is available that does away with the need for sand completely. These instant sandbags use water, and once soaked for a few minutes, an instant sandbag will fill up and act exactly the same way as a traditional sand bag.

Predictions and projections

Predictions and Projections of Flooding

Flooding is a growing threat to communities around the world, creating extensive damage to both property and lives. To help prepare for the increasing risk of flooding, predictions and projections can be made to better understand the potential impacts of flooding.

Predictions of flooding involve forecasting what may happen in the future, such as the likelihood, extent, and severity of future flooding events. These predictions are based on current knowledge of the climate, weather, hydrology, and other environmental factors

- Projections of flooding are based on more long-term predictions, such as the effects of climate change on flooding. This can include predictions on how higher temperatures and changing rainfall patterns may impact flooding in different areas around the world. These projections are based on computer models and can help to identify areas that are at risk of flooding in the future

- By understanding the predictions and projections of flooding, communities can be better prepared for potential flooding events and can implement better adaptation strategies to help reduce the risk of flooding.

A Flood Insurance Primer - Why Are So Few Homeowners Insured?

Flood insurance was a hot topic in the wake of Gulf Coast hurricanes Katrina and Rita. The lesson taken away from those disasters from a flood insurance perspective was generally the right one - The Congressionally-mandated flood insurance program does not work. Not nearly enough people buy flood insurance - ironically, far fewer buy mandatory flood insurance than would if the market were allowed to educate the public and convince them to buy it. To understand why so many homeowners even in hurricane prone areas lack flood insurance, it's necessary to learn a little bit about how flood insurance works in America.

The who and what of federal flood insurance

The Federal Emergency Management Agency (FEMA) designates flood zones based on a number of factors, all boiling down to the chance property in the zone will suffer flood damage. Whether federally subsidized flood insurance will be required (under circumstances described below) depends on the flood zone the property is or will be located in.

The National Flood Insurance Program (NFIP) makes federally subsidized flood insurance available, including where mandatory. (The mechanics of how insurance can be legally "mandated" are covered below.) Because NFIP is a federal government program - and so, someone else's money, unsullied by a profit motive -- flood coverage is incredibly cheap.

Flood zones and what they mean (for insurance purposes)

There are three basic types of flood zones designated by FEMA, subdivided into several more detailed zones.

Moderate to Low Risk areas are designated by flood zones B, C and X.

- Generally a less than 1% chance of flooding per year.
- Flood insurance is "available" to homeowners in these zones through the NFIP.

High Risk areas are designated by flood zones A, AE, A1-A30, AH, AO, AR and A99.

- Generally a greater than 1% chance of flooding per year.
- This generally translates into a 26% chance of flooding over the life of a 30-year mortgage.
- Mandatory flood insurance rules apply for mortgages in these zones.

High Risk - Coastal Areas designated by flood zones V, VE and V1-V30.

- Generally the same chance of flooding as A (High Risk) zones.
- Mandatory flood insurance rules apply for mortgages in these zones.

There is also a Zone D, "undetermined" risk area.

The gulf coast is almost entirely designated High Risk - Coastal Area.

"Mandatory" flood insurance

To understand what "mandatory" means when it comes to flood insurance, it's useful to step back and consider what Congress is and is not authorized to do under the Constitution.

The federal government cannot constitutionally mandate that people buy flood insurance. It cannot enforce building codes that would restrict the kind of construction authorized in certain flood zones.

What it can do is create a program, like the NFIP, and make it available to communities that pass and enforce flood zone building codes. You may be more familiar with Congress' threat to withhold highway funds to states that did not set a 55 and then 65 MPH speed limit. Same principle: What Congress cannot constitutionally require, it may accomplish by creating a benefit and threatening to withhold it.

So: Communities become eligible to participate in NFIP by taking steps to ensure new construction and existing structures mitigate flood risk.

NFIP was created in 1968 as a voluntary program. Because of low participation, Congress "mandated" (we're still getting to what that means) flood insurance in certain areas (now flood zones) in 1973. Participation remained low.

In 1994, Congress enacted flood insurance reform, continuing the "mandatory" nature of flood insurance and establishing new, severe sanctions for nonparticipation, in the form of requiring that homeowners having received relief purchase flood insurance to be eligible for similar help in the future.

You could stop reading here and know a lot about what's wrong with flood insurance: Congress said that it would only take care of uninsured homeowners' flood damage once. What this means to most people smart enough to have bought a home is that the federal government will take care of uninsured homeowners' flood damage once.

Who is subject to the "mandatory" flood insurance law?

Not the homeowner - rather, federally regulated lenders, GSEs and public agencies. These entities are required to ensure that any mortgage secured by structures in a flood hazard area has flood insurance.

If required, flood insurance will be required at the time a loan, including a refi, is made. Generally, notice is given to homeowners that they are required to purchase flood insurance at their expense. If they fail after notice, the lender may purchase it for them and add the cost to the monthly payment if the property is in a flood hazard area.

Life of loan monitoring is not required by law. (This becomes important in a way we will see.)

Lenders face civil money penalties -- no more than $100,000 aggregate per year -- if (and only if) they engage in a pattern or practice of shirking their flood insurance responsibilities.

Why might a homeowner in a flood-prone area not have insurance?

This is the heart of the matter. Considering the history, politics and division of responsibility for ensuring that flood-prone homeowners have insurance, here is why they don't:

1. People think homeowner's insurance covers floods. It doesn't.

2. Their property may not technically be in a flood zone designated by FEMA as requiring insurance, so it's not mandatory.

3. They worked through a non-federally regulated mortgage lender, that did not sell their loan to Fannie Mae or Freddie Mac, so it's not mandatory.

4. They have no mortgage -- it may be paid off or never have been encumbered (the 90-year-old home that's been in the family for three generations).

5. Lenders may not comply. A company originating $50 billion in mortgage loans in a quarter might economically view avoiding a possible $100,000 penalty as not worth the cost of rigorous compliance.

6. Homeowners get the insurance to get through closing, but then let coverage lapse, and they haven't been "caught" because there is no mandatory life of loan monitoring.

7. Their community may not participate in the program.

8. They assume the government will make them whole after losses without their buying insurance. Generally, they're right.

Flood insurance represents a failure of central planning and an apt demonstration of it inferiority to the free market. To better ensure that homeowners in hurricane prone areas are insured in greater numbers, Congress should bite the bullet and withhold aid where flood insurance was cheaply available and a choice was made not to purchase it (continuing to help those who lack insurance for reasons beyond their control). It should continue to require flood insurance at loan closing where it has the power to do so, but open the market to private insurance companies and require life-of-loan monitoring if it's serious about enforcing an insurance requirement. And penalties must be increased - the current one simply is not an economically feasible deterrent.

The role of policy and technology

The Role of Technology in Disaster Management

Although the short- and long-term consequences of increasingly frequent flooding vary from one community to the next, what is clear is that authorities across multiple levels of government need to invest in smart, strategic flood-response technology to prepare disaster response teams for the immediate aftermath of these events. The most effective flood response technologies range from drones for surveillance to situational awareness platforms that ensure valuable data reaches the right people. These tools allow decision-makers to build a comprehensive plan of action and respond to flooding events in real-time.

Although natural disasters have always presented considerable challenges to authorities – especially at the local level where resources and personnel can be limited – flood responses in particular demand a significant degree of coordination and collaboration. By incorporating surveillance, connectivity, and situational awareness technologies, it should be possible to launch emergency responses that are faster, better organized, and ultimately more effective in mitigating the damage of flood events.

It is essential that emergency response teams deployed to flooded areas operate from a shared common operational picture (COP) – one that provides reliable, real-time information as to the whereabouts of citizens, response crews, and infrastructural assets. In other words, the COP should enable high-fidelity situational awareness across all relevant personnel. By working from sound intelligence, disaster management efforts can be organized from the top down, rather than by way of the traditional "every crew for itself" mentality. At the same time, flood-response technology can enable and encourage personnel in the field to share important information with other deployed personnel and with a central headquarters.

Emerging Flood-Response Technologies

Recent floods have underscored the importance of communication and coordination in developing real-time responses. Fortunately, the technologies that are emerging are imminently capable of adding tremendous value on a number of fronts.

Pivotal flood-response technologies include:

- **Surveillance Technology** – Drones or unmanned aerial vehicles (UAVs) provide ground crews with an "eye in the sky" that allows them to gain a full picture of the flooding, including which areas are most affected and where emergency responses may be required next. Underwater drones can also help responders examine infrastructure and coordinate rescue efforts in heavily flooded areas.

- **Flood Mapping Technology** – Flood forecast maps use remote sensors to determine which areas are most at risk of flooding based on elevation, proximity to bodies of water, and other topographical data. They can also be helpful in evaluating if and when to rebuild infrastructure after a natural disaster, as some areas may have become too dangerous to accommodate homes and businesses.

- **Connectivity Technology** – The most powerful tool available during a natural disaster is one that nearly everyone has access to: a Smartphone. Quickly deployable cellular data communication platforms can help people stay in contact with their loved ones during floods. These networks can also help authorities communicate more easily with one another and with imperiled communities over social media.

> **Situational Awareness Technology** – Situational awareness platforms integrate discrete technologies, synthesize information streams, and activate data from UAVs, intelligent infrastructure, meteorological data, and more. Such integration helps responders build a COP, enabling crews on the ground to execute their critical responsibilities with far greater effectiveness.

A new generation of digital technologies is helping governments take decisive control over major flooding events. Perhaps the most important development in flood-response technology is the rise of situational awareness platforms. This technology enables decision-makers to effectively coordinate response efforts at a moment's notice, rather than executing strategies designed for dynamic situations that will almost certainly have changed by the time first responders arrive on the scene.

Integrating flood-response technologies and disaster response personnel into a situational awareness platform can make real-time coordination a reality, and help prioritize and distribute mission-critical information to the right people at the right time. It is even possible to leverage the power of crowd sourcing to pull citizen-generated data from social media and purpose-built public applications into a COP. Through the use of mobile data communication platforms, citizens are able to support the COP and receive vital information from government responders.

Ultimately, surveillance, connectivity, and situational awareness technologies have the potential to revolutionize how governments respond to major flooding events. By leveraging information in a more coordinated fashion and pulling from a wide array of assets across disaster-stricken areas, it is possible to develop responses that are better organized in turn – and save lives in the process.

> ### How technology can fight climate change

Solutions for the Internet of Things (IoT) with artificial intelligence (AIoT) are essential for addressing some of the problems related to carbon control. To increase the effectiveness, efficiency, and transparency of carbon management, three key areas are being concentrated on.

1. Integration of AIoT into reporting and measurement

The amount of work needed to just classify and organize the data from numerous business units and assets is enormous because there are so many databases and systems connected to various carbon-producing assets. The seamless sourcing of real-time activity level data and asset inventory data from a range of systems is made possible by AIoT integration. This gives a company the ability to structure, gather, and report data accurately for emissions monitoring and measurement, which lowers total data collection efforts and improves data quality and report resolution.

2. Abatement intelligence – predictive analytics to simulate emissions over time

Planning for abatement is difficult in part because there aren't any reliable methods for calculating the emissions produced by certain procedures. This problem is addressed by AIoT technology, which develops insights from real-time data to more accurately estimate process emissions. AIoT can enhance emissions estimates and improve performance evaluation of abatement systems by analyzing and learning from data from numerous operations. This technology not only optimizes abatement strategies but also minimizes the overall marginal abatement expenses.

3. **Carbon offsetting and offset integration**

With an anticipated potential market size of $200 billion by 2050, the carbon offset market, though a last resort, is crucial to reaching the worldwide net-zero emissions targets for nations and organizations. However, the business is plagued by issues with trading and certification of carbon offsetting. Technology can facilitate near-real-time REC validation and provide a market for quick and inexpensive carbon offsetting. Offset integration would give an organization access to a large pool of offsets from throughout the world, facilitating trade and emissions planning, easing administrative burdens, and optimizing the timing of REC purchases and retirement.

➢ Conclusion

Flooding is a natural disaster that can have devastating impacts on communities and individuals. It is caused by various factors, including heavy rainfall, melting snow, and the failure of man-made structures such as dams and levees.

The consequences of flooding can be widespread, including damage to property and infrastructure, loss of life, displacement of people, and economic losses. In addition, floods can cause environmental damage, such as soil erosion and the spread of diseases.

To minimize the impacts of flooding, it is important for individuals, communities, and governments to take proactive measures, such as creating and maintaining flood-resistant infrastructure, improving warning systems, and developing evacuation plans. Additionally, it is crucial to consider the impact of climate change on flood patterns and adjust mitigation strategies accordingly.

In conclusion, floods pose a significant threat to communities and individuals and require proactive and ongoing efforts to minimize their impacts.

It is also important to have effective recovery efforts in place to help communities and individuals rebuild and recover after a flood event. This can include providing financial assistance, supporting cleanup and rebuilding efforts, and addressing the long-term needs of affected communities.

Moreover, it is important to address the root causes of flooding, such as land use practices and urbanization, that increase the risk of flooding in some areas. Effective land use planning and development policies can help reduce the risk of flooding and protect communities from its impacts.

In conclusion, the impact of flooding can be reduced through a combination of mitigation measures, recovery efforts, and addressing the root causes of flooding. Collaboration between individuals, communities, and governments is essential in order to effectively manage and reduce the impact of flooding.

There are steps that can be taken, however, that have a high probability of success, and will have a positive effect of the economy. First, nuclear plant construction should be supported wherever safety can be achieved as a replacement for fossil fuel plants. Deep thermal well generator development and construction should also be supported as a replacement for fixed fossil fuel plants. Also, research for carbon dioxide sequestration should be supported as well as any other commercially viable alternate energy sources being developed now.

The possibility of climate change due to carbon dioxide emission is controversial. Many are not convinced that it exists. It cannot be ignored, however. If it is ignored, there is a possibility that it will turn into runaway global warming because of the thawing and decay of permafrost vegetation, and the loss of the reflective ice caps.

- **Call to action**

A Call to Action for Floods

Floods have become an increasingly devastating problem for communities around the world, resulting in extensive damage to both property and lives. In order to reduce the damage caused by floods, a call to action is needed to implement better adaptation strategies.

The first step is to educate people on the risks of flooding and on the various adaptation strategies that can be used to reduce the impact of flooding. This includes raising awareness of the potential impacts of flooding, such as the damaging effects it can have on homes, businesses, and infrastructure. Additionally, it is important to educate people on the various adaptation measures that can be taken to reduce the risk of flooding, such as developing better forecasting and early warning systems and enforcing land-use policies

The second step is to implement better adaptation strategies in communities. This requires collaboration between all stakeholders, such as local authorities, businesses, and citizens. It is also essential to make sure that there is an adequate budget for adaptation measures, such as developing flood defenses, restoring wetlands, and creating buffers

By implementing these steps, communities can be better prepared for potential flooding events and can reduce the damage caused by floods. It is therefore essential to take action now to ensure a better future for all.

The third step is to ensure that adaptation strategies are efficient and effective. This requires regularly monitoring the strategies to assess whether they are still relevant and are achieving the desired results. It is also essential to evaluate the performance of the strategies and to identify any areas that need improvement or further investment.

Finally, it is important to ensure that the adaptation strategies are sustainable in the long-term. This involves making sure that the strategies are resilient to potential changes in climate, weather, and other environmental factors. Additionally, it is important to keep up with the latest research and technological developments in relation to flooding, in order to ensure that the strategies are up to date and effective.

H. References and Further Reading

References:-

(i) Todays Zaman. Klaus Jurgens. The limitations of urban development: Have we reached the limitations of urban planning?

(ii) The Associated Programme on Flood Management (APFM).

(iii) Flash Flood Management in Urbanizing China, Xu Jianchu and Li Zhuoqing, Kunming Institute of Botany, the Chinese Academy of Science.

(iv) Prediction and management of flash floods in urban areas (URBAS), Thomas Einfalt , Andreas Wagner, Fritz Hatzfeld, Jörg Seltmann.

(v) IRIN, West Africa, Urban surge feeds flooding, Dakar, 14 September 2009.

John Dames is chief technology officer for Cool fire Solutions, a software company specializing in platform development and technology to deliver actionable intelligence. He has spent the past 8 years helping conceive and develop solutions for customers such as Enterprise Rent-A-Car, U.S. Military Special Forces, and municipal public safety and security teams.

- **Suggestions for additional reading.**

Facts About Flood Insurance and What You Need to Know in 2013

Damage from a flood is NOT covered on your home insurance policy!

Facts about floods:

- Floods and flash floods are the most common natural disaster, occurring in all fifty states
- One third of all flood insurance claims are generated outside areas considered flood-prone.
- Just a couple inches of water can do thousands of dollars of damage to a home
- One in five adults were not sure whether flood damage was covered in their standard homeowners policy You do not have to live near water to suffer a

flood loss

- Floods can be caused by many things such as heavy rain, melting snow, inadequate or overloaded drainage systems, dam or levee failure, hurricanes and more.

- Every year, flooding causes more then $2 billion of property damage in the U.S.

- Everyone lives in a flood zone

- Your home has a 26% chance of flooding as opposed to the 9% chance of fire during the course of a typical 30 year mortgage

- A flood is defined by FEMA as - A general and temporary condition of partial or complete inundation of two or more acres of normally dry land area or of two or more properties (at least one of which is your property) from:

- Overflow of inland or tidal waters,

- Unusual and rapid accumulation or runoff of surface waters from any source, or

- A mudflow; defined as A river of liquid and flowing mud on the surfaces of normally dry land areas, as when earth is carried by a current of water or

- The collapse or subsidence of land along the shore or a lake or similar body of water as a result of erosion or undermining caused by waves or currents of water exceeding anticipated cyclical levels that result in a flood as defined above.

What is flood insurance?

Flood insurance is a special policy backed by the National Flood Insurance Program (NFIP). Flood insurance covers the rising of flood waters from the ground. It does not cover broken pipes, or main breaks that may damage your home unless they are caused by flooding. It is important to note that homeowners insurance is designed to bring your home and its contents back to the same condition it was in before a loss, while flood insurance is only meant to get you back on your feet.

Who needs flood insurance?

According to FEMA - everyone. You might think that if you don't live in a beach-front property that you don't need flood insurance. However, beach-front properties only account for 3% of all flood losses. Everyone lives in a flood zone and flood damage can happen to your home even if you do not live near water. More than 25% of all flood claims come from low to moderate risk areas.

How do I purchase flood insurance?

First, check if your community participates in the program. Then contact your local insurance agent. Your agent will be able to sell you a policy through the NFIP. Don't wait to buy insurance until a flood is predicted- there is a 30 day waiting period for the insurance to go into effect.

Types of Emergencies - Floods
Floods Are the Deadliest Disasters

A great many of the largest disasters in recorded history have been floods. Whether biblical in nature such as Noah's flood or the devastation of Galveston by the 1900 hurricane known as "Isaac's Storm" (Named after the meteorologist who attempted to warn people of the storm) or Hurricanes Camille, Andrew and Katrina. All of these caused massive damage and death. Creating a plan to avoid a flooding disaster on these levels is pretty simple... get out before they happen.

Floods from dams bursting, river floods and storm surges from tropical storms, hurricane and typhoons have literally killed millions of people. China leads the list of deaths by flooding with a single flood in 1931 causing an estimated 4,000,000 deaths.

The single largest single day death toll in the US is from the storm that hit Galveston, TX in 1900 known as Isaac's Storm. The death toll was estimated as high as 12,000 souls. The majority of these were from the storm surge of 15 feet that engulfed the island that was only 8 feet above sea level.

The devastation was complete. No one who lived on the island went without loss of some kind.

Are You in a Flood Zone?

Do you live next to a stream, creek or river? If so, you are in a flood zone. Do you live downstream from a dam? Then you certainly are in a flood zone. Do you live near the beach? Then you are in a flood zone. Even if you live inland you can still be in a flood zone. What could be worse, you could live near a flash flood zone.

Contact FEMA to learn more about your location. You may be in a flood zone and not know it. Use this link to learn more about FEMA and even download an app to give you early warning alerts.

Flood Planning

This could be a simple as riding out the inconvenience of flooded roads and washed out bridges or complete evacuation due to your home being underwater or swept away. More than other types of emergencies, planning for a flood is specific to your location and surroundings.

For us, we live on high ground that has very little chance of a flood. There is virtually no chance of a flash flood hitting our home. Street flooding is not a threat as our home is up an 8 foot driveway and across the street is downhill to a series of ponds that can hold millions of gallons before flowing downhill to a major highway. We built in a no-flood zone that does not have a threat of even a '100 year' flood.

Most of the country is not situated as we are. The majority of the population lives near the coast. A rise in the seas by even a small amount could mean disaster as not only would the coast line flood but streams and rivers would back up, causing major flooding inland.

If you live nowhere near a flood zone then you should prepare for the inconvenience of being without electricity, food, medical care and water and no way to get any. You may be under a boil water order and need to filter your drinking water. If you are without power you will need a way to prepare food or have food that does not require cooking.

For flooding of your home plan on taking what you need to live on when you evacuate and move everything else up stairs before you go. If you do not have an upstairs take pictures of your belongings for insurance purposes.

Floods are Serious

Do not think you can ride out a flood. Floods are more serious than many people understand. Quite simply, floods kill people, destroy property and change landscapes forever. Know your situation. Analyze your surrounding and understand how high water will affect you.

Plan now for everything from an inconvenient flood all the way to full evacuation. This type of planning might save you life and the lives of your family.

How to Tell If Your Property Is in a Flood Zone

Living in a flood zone ups a person's odds of acquiring damage due to rising waters. For instance, most people residing in these areas will be affected by losses at some point, and those losses might be moved to the rest of society through systems for example publicly financed disaster relief. I know people who have been residing in a flood zone for 61 years and have not had flood damage. People living in a flood area or alongside a coastline must without a doubt evacuate. It's exactly the same as residing in a flood zone where by people pay a premium for an insurance plan.

Flooding in low-lying regions has been a dilemma for several years. This disaster has really an important policy problem for some towns. Some towns has carried out important efforts during the last 20 years to guarantee the safety of their open areas along with its developed neighborhoods. Flooding is undoubtedly an act of nature which has no regards to the damage to your house or personal things. Flooding of houses is a huge dilemma in most tropical areas in which there are higher volumes of rainfall. Individuals who reside in a region vulnerable to flooding are most likely aware about flood insurance as well as the high costs involved with living in a flood zone.

Mountain regions of big rivers and rivers that drain towards the coast frequently see rapid oncoming flooding. Land areas which might be at high risk for floods are known as Special Flood Hazard Areas, or flood plains. While damage research continues to be carried out in other places it's not certain that your region is going to be getting any federal help. Flood areas are land regions recognized by the Federal Emergency Management Agency (FEMA).

Flood insurance is usually excluded from most homeowner's insurance coverage. Flood insurance coverage is required if you reside in a flood zone, but this insurance policy is a great idea for virtually any property owner. Flood insurance is very important for individuals living in a flood area no matter if they own a house or private a small venture. Flood insurance will protect your property in the event of a flood. Flood insurance is accessible to property owners, tenants and businesses owners. Flood insurance coverage is necessary for anybody who lives in a flood region or where there is a possibility of a flood because of weather conditions. Flood coverage is a necessity for just about any mortgage lender if you reside in a special flood area. Flood insurance is accessible to homeowners, renters, condo owners/renters, and commercial owners or renters.

Are You At Risk of Flooding From Excess Rainfall?

Recently many areas of our state (and others as well) began the process of updating the FEMA Flood Insurance Rate Maps (FIRMs). These maps show the areas of potential flooding based on the 1-percent chance storm event. This has been known in the past as the 100 year flood and is also known as the Special Flood Hazard

Area (SFHA).

When you get the amount of rain comprising the 1-percent storm the flood water will come to a certain elevation near your home, known as the Base Flood Elevation (BFE). The FIRMs were recently updated for many counties around the country because of stimulus funding. FEMA is required to assess its flood hazard map inventory at least once every 5 years. But, because of funding shortfalls, it has been over 15 years for some communities.

For those homeowners with a mortgage, purchasing flood insurance is mandatory in a participating community if the loan is federally insured or the lender is regulated by the federal government. Flood insurance is highly advisable even if you're not required to purchase but are located near a stream or lake.

Remember, the 1-percent chance storm has a 1 percent chance of being met OR EXCEEDED in any year. Over the life of a 30 year mortgage there is a 26% chance of having a flood event that exceeds the base flood elevation. Mortgage insurance rates are generally less the higher above the base flood elevation your finished floor is located. Therefore, if you are four feet above the BFE the rates should be lower than if you were at or below the BFE. A $300 policy may well be worth the peace of mind it brings. Your homeowner's insurance policy has an exclusion from any flood damage.

You should also know that just because you're above the BFE and far away from a running stream, many dry ditches have caused significant damage to a home during a flash flood. Again, your homeowner policy is useless in this case but a flood policy would cover this damage.

"Purchasing flood insurance is mandatory...if the loan is federally insured or the lender is regulated by the federal government"

As stated above, your mortgage company may be required to ask you to purchase flood insurance. Of course, they would want you to do so because they are protected also. You should also know that the mortgage lender may also require flood insurance even if it is determined you don't need it. This is their prerogative. Again, the rates should be rather low in this case, but there are some costs nonetheless. Now that you know a little about the overall situation, how does this affect you directly? If you are currently shown to be in or near a flood hazard zone, or if you're going to be in or near a flood hazard on the proposed maps, NOW is the time to act. The following are the possible situations in which you may find yourself:

- A. Out of the flood hazard zone completely on the old and new maps. This is great. In this case there is no requirement for the purchase of flood insurance. But, as we said below, if there is ANY risk you might want to consider it. An evaluation of your risk is quick and easy.

- B. You're lot is currently or proposed to be shown in the flood hazard zone. This puts you under the requirement for flood insurance. Your situation may now be one of the following:

- C. Your lot is "in" the flood hazard zone but the lowest adjacent grade (LAG) around your house is "out" or above the base flood elevation (BFE). In this situation, it is possible that the flood insurance requirement may be removed. This process is called a Letter of Map Amendment (LOMA).

D. Your lowest adjacent grade (LAG) is below the BFE but the lowest finished floor elevation (FFE) is above the BFE. In this case you need to purchase flood insurance. An Elevation Certificate is necessary as a way to determine your premium rate.

E. Your lowest finished floor elevation (FFE) is below the BFE. This case is similar to 2.b. above but the flood risk is higher. Again, get an elevation certificate to determine your premium rate.

If you're in situation 2 above, the first step is to get an elevation determination. This process is done by a licensed land surveyor who will measure the elevation of your finished floor elevation and the lowest adjacent grade to determine your location relative to the flood hazard zone. This process will produce an Elevation Certificate that can be used to either complete the LOMA process or allow your insurance agent to set the flood insurance premium rate.

Urban Risks - Flash Floods

Disasters in their aftermath bring forth rising criticism leveled at those responsible for not only managing risks and disaster prevention, but also to the establishment accountable for urban planning. When disasters are deadly, with significant number of deaths involved, and stark pictures of horror, suffering, and loss widely exposed by the media - then, the criticism will increase to a crescendo of "poor management and negligence". The deadly flash floods rang the alarms for town planners to heed and pay more attention to environmental standards in rapid urbanization and industrialization. The full breadth of environmental realities and the state of natural resources together with steps which guarantees human safety should have provided the over-arching framework in making t decisions pertaining to transportation, industry and urban construction. (i)

Handling the challenges of flood risks in densely populated areas has been a constant historical factor in human settlements. Most cities are located in the valleys, flood plains and the coasts. Cities through their nature of having large impervious areas produce large run-offs which the drainage network cannot accommodate, and are potentially exposed to floods. It has been acknowledged that the damage potential of floods in the cities is extraordinarily high. Given the high population density in urban areas, even small scale flash floods may cause considerable damage. At the extreme end of the disaster spectrum, urban flash floods can result in disasters that set back development drastically. With climate change and global warming resulting in increased frequency of flood s and their magnitude, continuing urbanization and disproportionate growth, the economic costs of flash floods will soar. Sustainable management of urban flood risks is becoming an increasingly challenging task for city/municipal authorities. (ii)

Flash floods are distinctly characterized by very swift rise and recession, associated with debris flows and landslides, occurring along channels and rivers with small drainage area. Their distinct features paint a stark picture. Flash floods happen suddenly, easily and frequently, are very destructive, and difficult to protect against. Of late, flash floods brought extremely destructive disasters e.g. the recent flash floods in Istanbul, Turkey. In most cases it involves a break in flood protection facility.

Rapid economic growth aggravates flash flood hazards. As new construction takes over arable land, and urban population density increases, infrastructural growth may not proceed in tandem. Growth in urbanization inevitably reduces vegetation, wetlands and other habitats for flood prevention.

The patterns of urban flash floods are almost identical in its force. Small streams, canals, channels, and drainage ditches become fast flowing dangerous rivers. Where the terrain is flat, primary and secondary roads are inundated with torrents of floods, streets and parking lots becoming rivers of moving water. As the connotations imply, flash floods rise rapidly within a few minutes or hours of heavy rainfall. As the water rises rapidly and moves swiftly, carrying cars, ripping trees from the ground, and even destroying roads and bridges.

Disaster risk reduction in identified potential flood prone areas need to focus on extent of exposure and vulnerability. Exposure of urban dwellers close to river streams including infrastructure (roads, bridges, dams, power houses) located in the same area requires greater attention. Vulnerability could be minimized through increasing preparedness by way of flash flood guidance, community awareness campaigns, early warning systems, and planned coordinated emergency procedures. The World Conference on Disaster Reduction held in Kobe in January 2005 called for the early warning system to be people centered, providing timely and reliable warning to the people at risk.

While generally natural occurrences, flash floods are increasingly the result of human activities or poorly designed infrastructure. Very few countries have flash flood management action plan. Among those that do, China stands out with severe penalties for negligence. Flash floods are frequent features in China with two-thirds of the Chinese territory being mountainous, the recurrent natural disasters compounded by monsoon climate, fragile mountain terrain and increasing human activity. The threat confronts a total of 74 million population exposed to flash flood hazards in the mountain region. In a period of four decades (1950-1990) a total of 225,000 died in floods in China. (iii) The action plan calls for approval of any new construction in urban planning contingent upon completion of a flash flood assessment.

Prior to approving construction projects, city/municipal authorities could examine conditions affecting the construction area. Best practices in the management of flash floods in urban areas warrant enhancing the disaster management chain and assuring these extend into urban planning. Among some authorities, the approach include helping municipalities prepare for climate change. (iv)

Elsewhere, like in West Africa, there's a growing awareness that "urban surge feeds flooding", if left unplanned and unorganized, that is. Dakar's suburb of Guediawave was a dry area 30 years ago.

Nowadays, it's a different story. The residents of this densely populated suburb endure floods every rainy season. (v) Explosive population growth, poor urban management, urban congestion, and indiscriminate building in green belt zones all add on to shortening the fuse for disaster. Overpopulation in northern Nigeria has people building homes on waterways, and natural drainage system becoming blocked by rubbish. Despite bans on construction in the Dakar "cap vert" wetlands, this flood prone area received waves of rural-urban migration in the wake of the 70s and 80s Sahel-wide drought. Now the region is full of buildings and roads which block natural waterways and basins.

Explosive population growth, poor urban management, urban congestion, and indiscriminate building in green belt zones all add on to shortening the fuse for disaster.

What is a Flash Flood and How Do I Prepare For One?

Flash floods are frightening and dangerous because they are sudden and strong. Since they usually start and end in less than six hours, it is nearly impossible to prepare for one. How, then, does one protect one's family from these frightening occurrences? The best defense you can have is knowledge: know what causes them, what surrounding areas are susceptible to flooding, and what to do if you find yourself in a flash flood situation.

What is a Flash Flood?

A flash flood is a rapid flooding of a low-lying area of land, usually near a river or a stream. Most of these events occur after heavy rain, often in areas that don't usually receive much rainwater. Flash floods may also flow when man-made or natural ice dams break. In very rare instances, volcanic activity may melt glaciers, triggering one. When the ground can't absorb the rain as fast as it is falling, a these floods will occur. Excess water then runs into streams and rivers and flows quickly downhill. Depending on geography, flash flooding can happen as far as thirty miles away from the original site of precipitation.

Flash floods are one of the most dangerous types of natural disasters. According to the National Weather Service, flash floods kill more people each year than lightning, tornadoes, or hurricanes. Surprisingly, the speed of these damaging events is not their most dangerous quality. Indeed, most flash flood-related deaths transpire when people underestimate their power. Even a flood of just two feet can be swift and powerful enough to carry away an SUV-sized vehicle. The majority of fatalities occur when people attempt to ford flooded areas in their vehicles. Many other deaths are attributed to collisions with hidden debris, such as branches or logs that are pushed along by the water. When it comes to avoiding a flash flood-related injury or death, the best advice comes from the US National Weather Service: "Turn Around, Don't Drown."

Flood Damage Preparation

Due to their sudden nature, it is challenging to prepare for a flash flood. Still, there are a few precautions you can take if you live in an area that is prone to flooding. The most commonly flooded areas are southern and eastern states like Texas, Louisiana, Florida, New Jersey, and South Carolina. Desert areas of the West, such as Nevada, are also hotspots, since arid soil can't hold much water.

The most obvious precaution to take against flash floods is to stay up-to-date on weather patterns and flood warnings. Do not take any chances; if there is a chance a flood may hit your home or business, get out immediately. You can prepare for flash floods before any warning is announced by storing valuables and sentimental items in high places. A flood may be powerful, but if you keep valuables high, there is a smaller chance of water damage. Finally, try to save pictures of your valuables, in case they do get damaged. Having pictures of your house, car, furniture, and other assets will help speed up the insurance claim process following any disaster. (By the way, you do have flood insurance, right?)

Flash floods are dangerous, unexpected, and unpredictable. They can take lives, destroy homes, and cause extensive damage. Fortunately, with a little preparation, a flash flood doesn't have to be devastating.

www.ingramcontent.com/pod-product-compliance
Lightning Source LLC
Chambersburg PA
CBHW070320220526
45465CB00013B/1766